Read-About® Geography

We Need Directions!

By Sarah De Capua

Consultant
Nanci R. Vargus, Ed.D.
Primary Multiage Teacher
Decatur Township Schools, Indianapolis, Indiana

Children's Press®
A Division of Scholastic Inc.
New York   Toronto   London   Auckland   Sydney
Mexico City   New Delhi   Hong Kong
Danbury, Connecticut

Designer: Herman Adler Design
Photo Researcher: Caroline Anderson
The photo on the cover shows a family looking at a map.

Library of Congress Cataloging-in-Publication Data

De Capua, Sarah
  We need directions! / by Sarah De Capua
    p. cm. — (Rookie read-about geography)
Summary: An introduction to the cardinal directions and how to find them
using a compass, the Sun, and a map.
    ISBN 0-516-22574-X (lib. bdg.)   0-516-27380-9 (pbk.)
    1. Cardinal Points—Juvenile literature. [1. Cardinal points.
2. Orientation.] I. Title. II. Series.
    G108.5.C3 D4   2002
    912'.01'4—dc21
                           2001006641

Have you ever ridden in a car with adults who were lost?

They may have used a map
to find out which way to go.

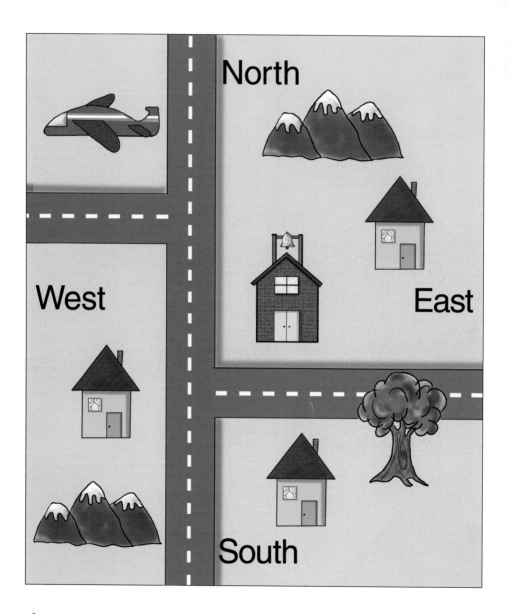

North

West

East

South

Maps show the directions
North, South, East,
and West. They are
the four major directions.

North is at the top of
a map. South is at the
bottom. East and West
are on the sides.

Look at a map or a
globe. Can you find
the compass rose?

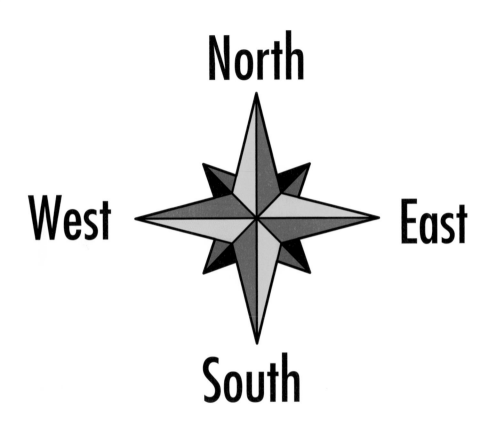

A compass rose points North, South, East, and West.

Look at the compass
rose on this map.

Which direction would
you go to get from the
school to the library?

If you said North,
you are right!

11

In what direction would you go to get from the houses to the park?

If you said East, you are right again!

You do not need a map
or a globe to find the four
major directions. You can
find them all by yourself.

15

Watch the Sun rise in the
morning. The part of the
sky where it rises is the East.

Watch the Sun go down in the evening. The Sun always sets in the West.

Stand with your right hand pointing toward the East (where the Sun rises). Point your left hand toward the West (where the Sun sets).

You are facing North.

North

West

East

South

North

West   East

South

Turn around. Now, point your left hand toward the East. Point your right hand toward the West.

You are facing South.

Imagine you are lost. You do not have a map. You do not know where East or West is. What can you do?

You can use a compass.
The arrow on a compass
always points North.

Hikers and backpackers use
compasses to find their way.

Some cars have built-in compasses that show the direction the car is traveling.

Look at a map of your town. Find the street where you live. In what direction is your school?

Which direction do you go in
to get to the police station?

Now that you know
the four major directions,
you can find your way
almost anyplace.

29

Words You Know

compass

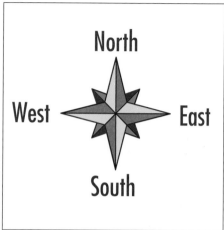

North

West — East

South

compass rose

East

globe

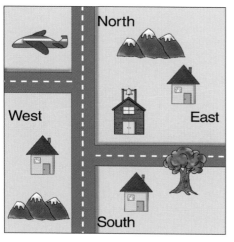

map

North

South

West

Index

About the Author

Sarah De Capua is an author and editor of children's books. She resides in Colorado.

Photo Credits